Letícia Winke Dias
André Pich Brunes
Andrea B. N. Martins

Maturation and phytochemical compounds in vegetables

Letícia Winke Dias
André Pich Brunes
Andrea B. N. Martins

Maturation and phytochemical compounds in vegetables

Volume I

ScienciaScripts

Imprint
Any brand names and product names mentioned in this book are subject to trademark, brand or patent protection and are trademarks or registered trademarks of their respective holders. The use of brand names, product names, common names, trade names, product descriptions etc. even without a particular marking in this work is in no way to be construed to mean that such names may be regarded as unrestricted in respect of trademark and brand protection legislation and could thus be used by anyone.

Cover image: www.ingimage.com

This book is a translation from the original published under ISBN 978-620-2-03247-6.

Publisher:
Sciencia Scripts
is a trademark of
Dodo Books Indian Ocean Ltd. and OmniScriptum S.R.L publishing group

120 High Road, East Finchley, London, N2 9ED, United Kingdom
Str. Armeneasca 28/1, office 1, Chisinau MD-2012, Republic of Moldova, Europe
Managing Directors: Ieva Konstantinova, Victoria Ursu
info@omniscriptum.com

Printed at: see last page
ISBN: 978-620-6-77622-2

SUMMARY

Summary

The production chain has been striving to perfect cultivation systems in order to guarantee the technical and economic success of its agricultural activities in the context of an increasingly demanding and competitive market. Some horticultural species are of great importance due to their high added value. In view of the importance these crops represent for Brazilian agriculture, it is necessary to expand technical and scientific knowledge in order to guarantee a quality product.

CHAPTER 1

Tomato fruit ripening

Letícia Winke Dias

Anderson Severo da Silva

David Bandeira da Cruz

The tomato plant belongs to the Solanaceae family, Solanum genus, Solanum lycopersicum species. It originated in the Andean regions of Peru, Bolivia and Ecuador and is one of the main vegetables produced in Brazil (NAIKA et al., 2006). It produces fleshy, indehiscent, juicy fruits of varying colour, known as berries, derived from a single ovary of a flower and containing numerous seeds (CHITARRA and CHITARRA, 2005).

Used as a model for the study of fleshy fruits, it is a diploid species with a relatively small genome of approximately 900 Mb (THE TOMATO GENOME CONSORTIUM, 2012). The advantage of using this species lies in its ease of cultivation (3 to 4 generations per year), propagation and genetic transformation, facilitating gene expression studies (GIOVANNONI, 2004). Tomatoes have thus been used as a model fruit, although the most studied plant species is still Arabidopsis thaliana, a very important and representative plant for various studies, although it does not produce fruit (PIRRELO, 2008).

Nutritionally, tomatoes are made up of approximately 94.5%

water and are a source of vitamins A, C and E, minerals such as potassium, phosphorus and calcium, carotenoids (lycopene and p-carotene) and phenolic compounds (DUMAS et al., 2003; CANENE-ADAMS et al., 2005). However, the composition varies according to the variety, stage of ripeness, light, temperature, soil, fertilisation, irrigation, and growing and storage conditions (YILMAZ, 2001; DUMAS et al., 2003).

Tomatoes are the second most economically important vegetable in the world (VIDOZ et al., 2010). World production of tomatoes in 2009 was 152.96 million tonnes, with China producing around 45.36 million tonnes, which represents 30% of the world's production, making it the country with the highest production of this commodity (FAO, 2009). Brazil's production in the same year was 4.31 million tonnes, making it the 9th largest producer and corresponding to around 3% of the world total (FAO, 2009).

Fruit ripeness can be determined by its colour, using colorimetry, the science of colour measurement that studies and quantifies how the human visual system perceives colour, in an attempt to specify it numerically since different stimuli are perceived in similar ways by observers (GUINÉ, 2007). The equipment used is the colorimeter, which uses sensors that simulate the way the human eye sees colour, and quantifies colour differences between a standard and a sample, always using the same light source and lighting method so that the measurement conditions do not change (BARROCA 2006).

When a colour is classified using a colorimeter, it is expressed in terms of: Hue - the range of wavelengths in which the colour is described; Brightness - the luminosity and; Saturation - the degree of purity (QUEVEDO, 2009).

Various systems are used to measure colour, the L* a* b* system or CIELab system is the most widely used and comprises three rectangular coordinates: L* measures the variation in luminosity between black (0) and white (100) corresponds to light and dark; a* is one of the chromaticity coordinates, and defines red colour for positive values and green colour for negative values; b* is the chromaticity coordinate, which defines yellow colour for positive values and blue colour for negative values, and also C* corresponds to purity (the stronger and brighter the colour, the further it is from the origin of the coordinates (Chroma): $C^* = (a^{*)(2)} + (b^{*)(2)}$ (QUEVEDO, 2009). According to Drlange (1994), the total colour difference can be calculated using the following formula:

$DE = (DL^{*)(2)} + (Da^{*)(2)} + (Db^*)$

In addition to the base coordinates, the rectangular coordinates, there are also the cylindrical coordinates:

H° corresponds to Hue

H° = arctg (b*/ a*), where a*>0 ; b*>0

$H° = 180° + arctg (b^*/ a^*)$, where $a^*<0$; $b^*>0$

$H° = 270° + arctg (b^*/ a^*)$, where $a^*<0$; $b^*<0$

$H° = 360° + arctg (b^*/ a^*)$, where $a^*>0$; $b^*<0$

In this context, tomatoes were assessed for their colour, harvested at two stages of ripeness up to the breaker stage.

Cherry tomatoes were sown and grown in a controlled house (phytotron®). The experiment was carried out in two stages of sowing and harvesting, with only fruit free of defects being harvested at the immature and mature stages.

The substrate used for cultivation was made up of soil with fertiliser, peat and perlite. It contained 40% dry material (raw products) and 60% organic material (mass of dry products). It had a pH in water of 6.0, electrical conductivity of 15 mS/m^{-1}, 70% water retention capacity and an NPK formulation of 14:16:18, respectively.

Fertigation was carried out every seven days, applying 2 $g.L^{-1}$ of NPK 16:6:16, corresponding to 16% nitrogen with 8.5% ammonia and 7.5% urea; 6% water-soluble phosphorus pentoxide and 16% water-soluble potassium oxide. Added magnesium and other micronutrients, formulated with 2% magnesium oxide; 0.01% total boron; 0.005% total copper chelated by EDTA; 0.1% total iron chelated by EDTA; 0.02% total manganese chelated by EDTA; 0.002% total molybdenum and; 0.005% total zinc chelated by

EDTA.

The tomato plants were grown in a controlled environment chamber (phytotron®), at a temperature of 25°C±2, relative humidity of approximately 75% and light flux - fluorescent lamps.

The colour of the fruit was measured using a colourimeter (Konica Minolta Chromometer Model CR 400/410, Japan) with an 8 mm aperture using the CIE-L*a*b* standard. To calculate the Hue angle (°Hue), which defines the colour tone, the a* and b* values were used (°Hue= $\tan^{-1}$b*/a*). °Hue values close to 180° represent greener fruit; those close to 90°, more yellow; and when they tend towards 0°, red. The measurements were taken on three opposite sides of each fruit.

The experiment was conducted using a completely randomised experimental design with four replicates. The colorimetry data was analysed for normality using the Shapiro-Wilks test and for homoscedasticity using the Hartley test. They were then subjected to analysis of variance ($p<0.05$). The comparison of means between treatments was assessed using the Tukey test ($p<0.05$).

The characterisations of the cherry tomato samples for the mean Hue angle ($À^0$) and Chroma showed that there was no interaction between the three factors for both variables. With regard to the °Hue variable, there was an interaction between harvest and stage, with only the main effect being observed for ripeness. With regard to chroma, there was an interaction between stage and harvest and between harvest and ripeness.

Fruits showed a higher Hue angle (°Hue) when they were in the green stage, both for fruits from batch 1 and batch 2. There was a reduction in the °Hue of fruits from batch 2 in the Breaker stage compared to batch 1, but these did not differ in the green stage. In general, mature fruit, regardless of access and stage, had a higher °Hue. The change in colour of tomato fruit from green to red, due to the degradation of chlorophylls and synthesis of lycopene, results in a reduction in the 'hue' angle and luminosity (REID, 2002).

Table 1: Hue angle (λ^0) in tomato fruit.

Ripening	Plots						
	1		Stadium	2			
	Green	Breaker	Average	Green	Breaker	Average	M. general
Mature	115,69	112,38	114,04	114,71	109,29	112,00	113,46 A
Immature	116,11	109,68	112.90	116,11	108,46	112.29	112.09 B
Average	115.90 a *a*	111.03 b *a*	113,47	115.41 a *a*	108.88 b *b*	112,14	
C.V. (%)				4,18			

Averages followed by the same lowercase letter in the row, uppercase in the column and in italics between accesses do not differ according to the Tukey test ($p < 0.05$).

For chroma, the highest results in batch 1 were observed in immature fruit and at the breaker stage, while they did not differ in relation to ripeness for batch 2. Again, chroma was highest at the breaker stage. There was a reduction in the chroma of fruit from batch 1 when they were in the breaker stage compared to fruit from batch 1 in this same stage.

Table 2. Chroma characterisation of tomato fruit.

Ripening	Access					
	2012			2013		
	Stadium					
	Green	Breaker	Average	Green	Breaker	Average
Mature	5,15	5,38	5.27 **a** B	5,19	6,08	5.64 **b** A
Immature	5,53	6,01	5.77 **a** A	5,24	6,17	5.71 **a** A
Average	5.34 b **a**	5.70 a **b**	5,52	5.22 b **a**	6.13 a **a**	5,67
C.V. (%)	18,46					

Averages followed by the same lowercase letter in the row, uppercase in the column and in italics between accessions do not differ according to Tukey's test ($p < 0.05$).

Fruits harvested when ripe and green in colour have a higher Hue angle. On the other hand, the highest colour is observed in tomato fruits at the breaker stage.

The colour and Hue angle of the fruit of the same tomato cultivar can vary according to the access from which they were obtained.

References

BARROCA, M.J.; GUINÉ, R.P.F.; PINTO, A.; GONÇALVES, F.M.; FERREIRA, D.M.S.F. **Food and Bioproducts Processing**, n.

84, p.109-113, 2006.

CANENE-ADAMS, K.; CAMPBELL, J.K.; ZARIPHEH, S.; JEFFERY, E. H.; ERDMAN, J. W. Jr. The Tomato as a Functional Food. **The Journal of Nutrition**, v. 135, p. 1226-1230, 2005.

CHITARRA, M.I.F.; CHITARRA, A.B. **Post-harvest of fruit and vegetables: physiology and handling**. 2 ed. Lavras: UFLA, 2005. 783p.

DRLANGE. Colour review. Drlange Application Report No. 8.0e.USA, 1994.

DUMAS, Y.; DADOMO, M.; DI LUCCA, G.; GROLIER, P. Effects of environmental factors and agricultural techniques on antioxidant content of tomatoes. **Journal of the Science of Food and Agriculture**, v. 83, p. 369-382, 2003.

FOOD AND AGRICULTURE ORGANISATION OF THE UNITED NATIONS. Tomato. Available at:< http://www.fao.org/nr/water/cropinfo_tomato.html> (2009).

GIOVANNONI, J.J. Genetic Regulation of fruit development and ripening. **The Plant Cell,** v. 16, p. 170-180, 2004.

GUINÉ, R.P.F.; FERREIRA, D.M.S.; BARROCA M.J.; GONÇALVES F.M. **Biosystems Engineering,** v.98, n.4, p.422-429, 2007.

NAIKA, S.; JEUDE, J.V.L.; GOFFAU, M.; HILMI, M.; VAN DAM, B.
Growing tomatoes: production, processing and marketing. 1ed. Wageningen: Agromisa Foundation and CTA, p.104, 2006.

PIRRELLO, J. **Caractérisation moléculaire et physiologique des Facteurs de Réponse à l'Ethylène (ERF) chez la tomate (Solanum lycopersicon).** 2008.(Doctoral Thesis) - l'Institut National Polytechnique de Toulouse. France.

QUEVEDO, R.; DIAZ, O.; CAQUEO, A.; RONCEROS, B.; AGUILERA, J.M. **LWT - Food Science and Technology**, v.42, p.1367-1373, 2009.

REID, M.S. **Maturation and Maturity Indices**. In: KADER, A.A. (E.d). Postharvest Technology of Horticultural Crops. 3, ed. University of California. Agriculture and Natural Resources. 2002. Publication 3311, p. 55-66, 2002.

THE TOMATO GENOME CONSORTIUM. The tomato genome

sequence provides insights into fleshy fruit evolution. **Nature**, v. 485, p. 635 - 641,2012.

VIDOZ, M. L.; LORETI, E.; MENSUALI, A.; ALPI, A.; PERATA, P. Hormonal interplay during adventitious root formation in flooded tomato plants. **The Plant Journal**. v. 63, p. 551-562. 2010.

YILMAZ, E. The chemistry of fresh tomato flavour. Turkish Journal **of** Agriculture **and Forestry**, v. 25, p. 149-155, 2001.

CHAPTER 2

VITAMIN C CONTENT AFTER HEAT TREATMENT IN HYBRID AND SPRIG BROCCOLI

Letícia Winke Dias

Cristiane Mariliz Stocker

Vitor Mateus Kolesny

Broccoli (Brassica oleracea L.) belongs to the Brassicaceae family and is one of the most consumed vegetables due to its content of vitamins, antioxidants and anticarcinogenic compounds (YUAN et al., 2010). Some authors have shown a 56 per cent association between increased consumption of broccoli and protection against cancer (VERHOEVEN et al., 1996). This protective effect is attributed to the phytochemicals present in this vegetable, which include vitamins C and E; flavonols, quercetin and kaempferol; and carotenoids, p-carotene and lutein (PODSEDEK, 2007). In general, brassica vegetables have higher levels of L-ascorbic acid and sulphur compounds than other vegetables.

Vitamin C is defined as a general term for compounds that exhibit the biological activity of L-ascorbic acid. Blocket et al. (2004) reported that this vitamin can reduce levels of C-reactive protein, a marker of inflammation and possible factor in coronary heart disease. It also works to prevent scurvy, maintain healthy skin and blood vessels, form collagen, absorb inorganic iron and reduce cholesterol (BURDULU, 2006). Vitamin C has also been shown to inhibit the formation of nitrosamines in vivo from nitrates and nitrites used as preservatives.

However, L-ascorbic acid is sensitive to destruction when the plant is subjected to adverse handling and storage conditions. Losses are accelerated by high temperatures, low relative humidity, physical damage and cold injury. Ascorbic acid is easily oxidised, especially in aqueous solutions, a process that is favoured by the presence of oxygen, heavy metal ions, especially Cu^{2+}, Ag^{+} and Fe^{3+}, alkaline pH and high temperature (PARVIAINEN and NYYSSONEN,1992).

Based on the above, the aim was to check the levels of L-ascorbic acid in broccoli with hybrid and branchy inflorescences, in fresh form and after heat treatment.

Broccoli from local shops in the city of Pelotas, RS, purchased in January 2017, was used for this study. The experimental design was completely randomised, with the treatments being the inflorescences, hybrid (single head) and branchy (lateral and numerous) and the application of heat treatments, blanching and canning, as well as the fresh product, which was used as a control. The blanching process was carried out at a temperature of 65°C for 15 minutes. In addition to blanching, vinegar, water, salt, pepper and bay leaf were also added to the preserves, after which the samples were packed, sealed hermetically and subjected to thermal processing and cooling (LEITÃO et al., 2009).

To determine the L-ascorbic acid content, 50 grams of the sample were crushed and homogenised, 10 grams were removed and 30 mL of a solution of metaphosphoric acid (4.5%) in ultrapure water was added, leaving the mixture to stand for 1 hour in a bottle protected from light. This volume was then transferred to a 50 mL volumetric flask and topped up with ultrapure water. The sample was filtered with cotton wool and centrifuged in eppendorff tubes at 7,000 rpm for 10 minutes. The supernatant was used to analyse L-ascorbic acid using high-performance liquid chromatography (HPLC).

The chromatographic analysis consisted of a Shimadzu HPLC system equipped with an automatic injector and UV detector (254 nm). The separation was carried out on a Shim-Pak CLC-ODS reversed phase analytical column (3.9cm x 150mm x 4mm), with as the stationary phase for octadecyl groups. Elution was carried out using a gradient system, with a flow rate of 0.8 mL min-1, developed with 100% 0.1% (v/v) aqueous acetic acid solution, changing linearly in 5 minutes to 98% (v/v) 0.1% (v/v) acetic acid solution and 2% (v/v) methanol, maintaining this proportion for up to 7 minutes and returning to the initial proportion in 10 minutes.

L-ascorbic acid was identified by comparing the retention times obtained with the standard analysed under the same conditions. Quantification was done by external standardisation and the results expressed in mg of L-ascorbic acid per 100 grams of fresh mass.

The data was analysed for normality using the Shapiro-Wilk test and for homoscedasticity using the Hartley test. Subsequently, they were subjected to analysis of variance ($p<0.05$). The effects of inflorescences were assessed using the t-test ($p<0.05$) and heat treatment using the Tukey test ($p<0.05$).

There was an interaction between the factors tested for the variable studied (Table 1). As for the inflorescences, there were differences in both the hybrid and the branch between fresh, blanched and preserved. When comparing the hybrid and the branch, both the in natura form and the blanching differed, but the canning showed no difference due to the lack of detection of L-ascorbic acid in this treatment.

The highest levels were found in broccoli with hybrid inflorescences in fresh form, followed by blanching. A reduction in ascorbic acid content of around 41 per cent in broccoli after blanching was reported by Lisiewska and Kmiecik (1996). Howard et al. (1999) found a reduction of around 30 per cent in the content of this vitamin in steamed broccoli. Franke et al. (2004) described values of between 41 and 64 mg $100g^{-1}$ of ascorbic acid in fresh broccoli and 18 to 21 mg $100g^{-1}$ in cooked broccoli, values very close to those found in this study (Table 1). The decrease in ascorbic acid content in broccoli was also reported by Zhang and Hamauzu (2004), who found around 103 mg $100g^{-1}$ in fresh broccoli and after cooking for 5 minutes, the value dropped to 35 mg $100g^{(-1)}$.

A large loss of L-ascorbic acid was observed after bleaching, 227% for the hybrid and 69% for the branchy type. Although the hybrid had a much higher initial L-ascorbic acid content, blanching reduced it to values close to those of the branchy type. In the case of tinned fruit, the loss was total (Table 1). This is possibly due to the fact that the broccoli was blanched in water. It is known that one of the factors that most influences the loss of vitamin C during food preparation is leaching. As vitamin C is water-soluble, during cooking in water part of the vitamin is lost to the water, which is generally discarded. The shorter the cooking time and the amount of water used, the greater the retention of ascorbic acid (HOWARD et al., 1999).

Table 1. L-ascorbic acid C content (mg $100g^{-1)}$ in broccoli with hybrid and branchy inflorescences in fresh form and under two forms of heat treatment.

Inflorescences	L-ascorbic acid (mg $100g^{-1}$)		
	In natura	Bleaching	Preserves
Hybrid	67,0* a[1/]	20,5* b	0.00ns c
Ramoso	28,0 a	16,6 b	0,00 c

[1/] Averages followed by **the** same lowercase letter in the line do not differ

by the t-test (p<0.05) comparing the effect of heat treatment within each inflorescence. ns, *, Non-significant and significant, respectively, by the **t-test (p<0.05) comparing** the effect of inflorescence within each heat treatment.

In this study, it was found that the L-ascorbic acid content varies between the types of inflorescences and that heat treatment caused a decrease in the concentration of this compound.

References

BLOCK, G., JENSEN, C., DIETRICH, M., NORKUS, E. P., HUDES, M., PACKER, L. Plasma C-reactive protein concentrations in active and passive mokers: Influence of antioxidant supplementation. J. Amer. Coll. **Nutrition,** v. 23, p. 141-147, 2004.

BURDULU, H.S.; KARADENIZ, F. Degradation of vitamin C in citrus juice concentrates during storage. **J. F. Engineering**, v.74, p.211-216, 2006.

FRANKE, A. A.; et al. Vitamin C and flavonoid levels of fruits and vegetables consumed in Hawaii. **J. Food Comp. Anal.**, v. 17, p. 135, 2004.

HOWARD, L.A.; WONG, A.D.; PERRY, A.K; KLEIN, B.P. (3-

carotene and ascorbic acid retention in fresh and processed vegetables. **J. Food Sci.**, v. 64, p. 929-936, 1999.

LEITÃO, A.M.; BORGES, C.D.; BARBOSA, E.G.; CHIM, J. F. **Frutas e hortaliças em conservas**. Pelotas: Ed. Universitária da UFPel, 2009. 78 p.

LISIEWSKA, Z.; KMIECIK, W. Effects of level of nitrogen fertilizer, processing conditions and period of storage of frozen broccoli and caulifl ower on vitamin C retention. **Food Chem.**, v. 57, n. 2, p. 261270, 1996.

PARVIAINEN, M.T.; NYYSSONEM, K. **Ascorbic acid**. In: Leenheer, A.P.D; Lambert, W.E. Nelis, H (eds) M. Chro. Ann. Vitamins. New York: Marcel Dekker, 1992.

PODSEDEK, A. Natural antioxidants and antioxidant capacity of Brassica vegetables: A review. **LWT-F. Sci. Technology**, v. 40, p. 1-11, 2007.

VERHOEVEN, D.T.H.; GOLDBOHM, R.A.; VAN POPPEL, G.; VERHAGEN, H.; VAN DEN BRANDT, P. A. Epidemiological studies on Brassica vegetables and cancer risk. C. **Epid. Bio. Prevention**, v. 5, p. 733-748, 1996.

YUAN, G.; SUN, B.; YUAN, J.; WANG, Q. Effect of 1-methylcyclopropene on shelf life, visual quality, antioxidant enzymes and health-promoting compounds in broccoli florets. **Food Chem.**, v. 118, p. 774-781, 2010.

ZHANG, D.; HAMAUZU, Y. Phenolics, ascorbic acid, carotenoids and antioxidant activity of broccoli and their changes during conventional and microwave cooking. **Food Chem.**, v.88, n. 9, p. 503-509, 2004.

CHAPTER 3

Chlorophyll and total carotenoid levels in broccoli

Letícia Winke Dias

Cristiane Mariliz Stocker

Scharlise Diovanella Schneider Da Silva

Broccoli (Brassica oleracea L.) is a widely consumed vegetable among Brazilians, mainly eaten fresh or minimally processed. It is estimated that per capita consumption of fruit and vegetables has grown by 25 per cent in the last decade (ROMBALDI et al., 2007), and that for broccoli this percentage has been even higher. This increase is probably due to the popularisation of the knowledge that broccoli is a vegetable that is a source of potentially protective phytochemical compounds, including phenolic compounds, carotenoids, selenium and vitamins (MANCHALI et al., 2012). In addition, its potential anticarcinogenic and antioxidant benefits have been attributed to its relatively high content of glycosinalates (LATTÉ et al., 2011).

The yellowing of vegetables, especially broccoli, is one of the most easily perceptible alterations after harvest, interfering with the commercial value of the product (ASODA et al., 2009). This change is accompanied by losses of bioactive components such as ascorbic acid (JEFFERY & ARAYA, 2009), glucosinolates and folic acid (JONES et al., 2010).

The action of the ethylene molecule, a hydrocarbon, gaseous hormone, which diffuses freely from one cell to another through the membranes acting as an activator of a cascade of signals related to the expression of genes linked to ripening and increased synthesis of chlorophyllases, cellulases, polygalacturonase, pectinamethylesterase, pyruvate dehydrogenase (DEL AGUILA, 2013; SIVAKUMAR & WALL, 2013). This coordinating function of ethylene is reinforced by its ability to stimulate its own synthesis in climacteric fruits known as autocatalytic synthesis, i.e. ethylene synthesis generates more ethylene, further increasing the ripening or yellowing process (PUN et al., 2013).

Vegetables have been highlighted as excellent sources of carbohydrates, vitamins, minerals, antioxidants and fibre. Scientific research is therefore beginning to focus on studies relating these foods to consumer health protection (EINBOND et al., 2004). Various methods have been used to extend the post-harvest life of intact broccoli heads, including refrigeration (TOIVONEN, 1998a); modified atmosphere (RAI et al., 2008), heat (COSTA et al., 2005a), UV-C radiation (BAPAT et al., 2010) and the use of 1-methylcyclopropene (1-MCP) (LEE et al., 2012).

The photosynthetic pigments present and their abundance vary according to the species. Chlorophyll a is present in all organisms that carry out photosynthesis and is the pigment used to carry out the photochemical phase, while the other pigments help to absorb light and transfer energy to the reaction centres, and are therefore called accessory pigments. Chlorophyll b, carotenoids and phycobilins are the so-called accessory pigments (TAIZ & ZEIGER, 2009).

In this context, the objective was to verify the changes in colour and in the levels of chlorophylls and total carotenoids in broccoli after storage.

The work was carried out using branched and hybrid broccoli cultivars. Broccoli from local shops in the city of Pelotas, RS, was used for this study. The experimental design was completely randomised, with the treatments consisting of the inflorescences, hybrid (in the shape of a single head) and ramosa (in the shape of a side), and storage applications: room temperature at 24°C ± 1, without the presence of light; refrigerated atmosphere with relative humidity of 80% ± 5, at 4°C ± 1; and freezing at -20°C, in addition to the control treatment (untreated).

Evaluations took place four and six days after the treatments were applied. The response variables evaluated were moisture content, colour, total chlorophyll, a, b and total carotenoids. The degree of humidity was determined by heating at 105 °C in a forced-air oven until constant weight was reached and the results were expressed as a percentage of dry matter (AOAC, 2005). The colour of the inflorescences was measured using a colourimeter (Minolta Chromometer Model CR 300, D65, Osaka, Japan) with an aperture of eight mm using the CIE-L*a*b* standard. To calculate the Hue angle (°Hue), which defines the colour tone, the a* and b* values were used (°Hue= tan-1b* /a*). °Hue values close to 180° represent greener inflorescences; those close to 90° are more yellow; and when they tend towards 0°, red. The measurements were taken on opposite sides of each inflorescence.

The chlorophyll and total carotenoid contents were determined using 1 g samples macerated in a mortar in the presence of 5 mL of 80% (v/v) acetone. The material was centrifuged at 10,000 rpm for 10 minutes and the supernatant was transferred to a 25 mL volumetric flask and topped up with 80% (v/v) acetone. The chlorophyll a, b, total (a+b) and total carotenoid contents were calculated using the equations (Equations 1, 2, 3 and 4) established by LICHTENTHALER (1987) from the absorbance of the solution obtained by spectrophotometry at 647, 663 and 470 nm. The results were expressed in mg.g-1 of dry mass (DM).

Total Chl = 7.15 (A663) + 18.71 (A647) (1)

Chi 'a'= 12.25 (A663) - 2.79 (A647) (2)

Chl 'b'= 21.50 (A647)- 5.10 (A663) (3)

Total carotenoids = [1000(A470)-1.82 Chla - 85.02 Chl b] /198 (4)

The data was analysed for normality using the Shapiro-Wilk test and for homoscedasticity using the Hartley test. Subsequently, they were subjected to analysis of variance (p<0.05). The effects of inflorescences were assessed using the t-test (p<0.05), storage using the Tukey test (p^0.05) and comparison with the control using the Dunnett test (p<0.05).

There was an interaction between the factors tested for all the variables studied. It was not possible to determine the analyses for the branchy inflorescence in the room temperature treatment, as the broccoli was not fit for consumption (Tables 1 and 2).

When studying the moisture content of the hybrid inflorescence, the storage treatments differed over time. When compared to the control, there was no significant difference only in relation to room temperature. In terms of °Hue, the refrigerated atmosphere and freezing treatments preserved the green colour when compared to room temperature, while the control treatment showed chlorophyll degradation (Table 1).

TABLE 1. Moisture on a dry basis (%) and °Hue angle in broccoli with hybrid and branchy inflorescences after different forms of storage.

Storage	Humidity		°Hue	
	Hybrid	Ramoso	Hybrid	Ramoso
Witness (T)	20,28	20,91	166,44	125,57
		4 DAT [2/]		
Ambient temperature	$18,83^{b}$	nd[3/]	$90,77^{a*}$	nd
Refrigerated	32.09^{aA*}	$20,19^{B}$	110.38^{aA}	$116,03^{A}$

atmosphere			*	
		6 DAT		
Ambient temperature	15,41 c *	nd	102,83 b *	nd
Refrigerated atmosphere	33.42 aA *	33.34 aA *	119.69 aA *	116.84 aA *
Freezing	30.97 bA *	26.56 bB *	123.86aA *	118.56aA *

· Significant by Dunnett's test ($p<0.05$) in relation to the control (T). 1/ Averages followed by the same lowercase letter in the column do not differ by Tukey's test ($p<0.05$) comparing the effect of storage within each inflorescence. Averages followed by the same capital letter in the row do not differ by the t-test (p-value 0.05) comparing the effect of inflorescences within storage. 2/ DAT: days after treatments. 3/ nd: not determined.

Table 2. Chlorophyll a, b, total (a+b) and total carotenoid contents (mg g^{-1} of DM) extracted from broccoli with hybrid and branchy inflorescences after different forms of storage. FAEM/UFPel,

Storage	Chlorophyll a		Clrofila b		Total chlorophyll		Total carotenoids	
	Hybrid	Ramoso	Hybrid	Ramoso	Hybrid	Ramoso	Hybrid	Ramoso
Witness (T)	0,064	0,144	0,015	0,199	0,079	0,343	12,98	25,49
				4 DAT2/				
T. environment	0.040 bA	nd 3/	0,060 b *	nd	0,100 b *	nd	15,90 a	nd
A. refrigerated	0.117 aB *	0,277 A *	0.205 aB *	0,483 A *	0.322 aB *	0.760 aA*	15.32 aB	66,43 A*
				6 DAT				
T. environment	0,176 a	nd	0,256 a *	nd	0,432 a *	nd	32,60 a *	nd
A. refrigerated	0.060 cB	0.197 aA *	0.016 bB	0.316 bA *	0.076 cB	0.513 bA*	8.69 bB *	53.21 bA*
Freezing	0.102 bB *	0.248 aA *	0.062 bB *	0.591 aA *	0.164 bB *	0.839 aA*	9.26 bB *	95.54 aA*

· Significant by Dunnett's test (p^0.05) in relation to the control (T). (1/) Averages followed by the same lowercase letter in the column do not differ by Tukey's test ($p<0.05$) comparing the effect of storage within each inflorescence. Averages followed by the same capital letter in the row do not differ by the t-test (p^0.05) comparing the effect of inflorescences within storage. 2/ DAT: days after treatments. (3/)nd: not determined.

However, TOIVONEN et al. (1998b) found that there are differences in the response of broccoli genotypes to yellowing. According to TENG & CHEN (1999), chlorophyll pigments are susceptible to chemical and physical changes during vegetable processing. Changes in colour as a result of heat treatment are mainly the result of the conversion of chlorophyll into pheophytin through the replacement of chlorophyll's Mg^{+2} with hydrogen

(KAISER et al., 2012). The speed of the pheophytinisation reaction is generally greater than that of other chlorophyll degradation routes, and it is considered the most important mechanism for destroying chlorophyll during the processing and storage of vegetables (MARTINS et al., 2002). According to data from COSTA et al. (2005b), the loss of green colour in broccoli occurs with a reduction in chlorophyll a and b content, resulting in a decrease in °Hue values, with the greatest reductions referring to chlorophyll a content.

The refrigerated and freezing atmospheres show the preservation of the green colour, as well as a significant increase in the chlorophyll a, chlorophyll b and carotenoid contents when compared to room temperature.

It was found that the degree of humidity varied between the types of inflorescences and throughout the application of the treatments. The refrigerated atmosphere and freezing treatments preserved the green colour when compared to room temperature. Chlorophyll a, chlorophyll b and carotenoid contents generally showed significant increases.

References

ABLE, A. J.; WONG, L.S.; PRASAD, A.; O'HARE, T.J. The effects of 1-methylcyclopropene on the shelf life of minimally processed leafy asian vegetables. **Postharvest Biology and**

Technology, v. 27, n. 2, p. 157-161, 2003.

ALI, A.; ONG, M.K.; FORNEY, C.F. Effect of ozone pre-conditioning on quality and antioxidant capacity of papaya fruit during ambient storage. **Food Chemistry**, v.142, p. 19-26, 2014.

ASSOCIATION OF OFFICIAL ANALYTICAL CHEMISTS (AOAC). 17 ª ed. **Official Methods of Analysis**. Arlington, 2005.

ASODA, T.; TERAI, H.; KATO, M.; SUZUKI, Y. Effects of postharvest ethanol vapour treatment on ethylene responsiveness in broccoli. **Postharvest Biology and Technology,** v. 52, p. 216-220, 2009.

BAPAT, V. A.; TRIVEDI, P. K.; GHOSH, A.; SANE, V. A. GANAPATHI, T. R., NATH, P. Ripening of fleshy fruit: Molecular insight and the role of ethylene. **Biotechnology Advances**, v. 28, p. 94-107, 2010.

COSTA, M.L.; CIVELLO, P.M.; CHAVES, A.R.; MARTINEZ, G.A. Effect of ethephon and 6 benzylaminopurine on chlorophyll degrading enzymes and a peroxidase - linked chlorophyll bleaching during postharvest senescence of broccoli (Brassicaoleracea L.) at

20°C. **Postharvest Biology and Technology**, v. 35, p. 191-199, 2005.

DEL AGUILA, J.S. **Hormones and plant growth regulators: Plant Morphology and Physiology**. Available at:<http://200.132.139.11/ aulas/Eno> Accessed on 19/01/2014.

EINBOND, L.S.; REYNERTSON, K.A.; LUO, X.; BASILE, M.J.; KENNELLY, E.J. Anthocyanin antioxidants from edible fruits. **Food Chemistry**, v. 84, p. 23-28, 2004.

JEFFERY E.H.; ARAYA M. Physiological effects of broccoli consumption. **Phytochemistry Reviews**, v.8, p. 283-298, 2009.

JONES R.B.; FRISINA C.L.; WINKLER S.; IMSIC M.; TOMKINS R.B. Cooking method significantly effects glucosinolate content and sulforaphane production in broccoli florets. **Food Chem.**, v.123, p.237-242, 2010.

KAISER, A.; BRINKMANN, M.; CARLE, R.; KAMMERER, D.R. Influence of thermal treatment on colour, enzyme activities, and antioxidant capacity of innovative paste like parsley products. **Journal of agricultural and food chemistry**, v.60, n.12, p. 32913301, 2012.

LATTÉ, K.P.; APPEL, K-E.; LAMPEN, A. Health benefits and possible risks of broccoli - An overview. **Food Chem Toxicol.**, v.49, n.12, p. 3287-3309, 2011.

LEE J.; CHENG L.; RUDELL D.R.; WATKINS C.B. Antioxidant metabolism of 1-methylcyclopropene (1-MCP) treated 'Empire' apples during controlled atmosphere storage. **Postharvest Biology and Tecnology**, v. 65, p. 79-91, 2012.

LICHTENTHALER, H.K. Chlorophylls and carotenoids: pigments of photosynthetic biomembranes. In: PACKER, L.,DOUCE, R. (Eds.). **Methods Inenzimology**, London: Academic Press, v.148, p. 350381, 1987.

MANCHALI, S.; MURTHY, K.N.C.; PATIL, B.S. Crucial facts about health benefits of popular cruciferous vegetables. **J Funct Foods**, v.4, n.1, p. 94 - 106, 2012.

MARTINS, R.C.; SILVA, C.L.M. Modelling colour and chlorophyll losses of frozen green beans (Phaseolus vulgaris, L.). **International Journal of Refrigeration**, v. 25, p. 966-974, 2002.

PUN, U.K.; NIKI, T.; ICHIMURA, K. Ethanol reduces sensitivity to ethylene and delays petal senescence in cut Tweedia caerulea flowers. **Plant Growth Regulation**, n.69, p. 125-1330, 2013.

RAI, D.R.; TYAGI, S.K.; JHA, S.N.; MOHAN, S. Qualitative

changes in the broccoli (Brassica oleracea italica) under modified atmosphere packaging in perforated polymeric film. **Journal of Food Science and Technology**, v.45, p. 247-250, 2008.

ROMBALDI, C.V.; TIBOLA, C.S.; FACHINELLO, J.C.; SILVA, J.A. Perception of consumers in Rio Grande do Sul in relation to

fruit quality requirements. **Revista Brasileira de Fruticultura**, v. 29, p. 681-684, 2007.

SERRANO, M.; MARTINEZ - ROMERO, D.; GUILLÉN, F.; CASTILLO, S.; VALERO,D. Maintenance of broccoli quality and functional properties during cold storage as affected by modified atmosphere packaging. **Postharvest Biology and Technology**, v. 39, p. 61 - 68, 2006.

SIVAKUMAR, D.; WALL, M.M. Papaya Fruit Quality Management during the Postharvest Supply Chain. **Food Reviews International**, v.29, n.1, p. 24-48, 2013.

TAIZ, L.; E ZEIGER, E. **Plant Physiology**. Sinauer, Assoc. Inc. Sunderland Ma, USA, 5ª ed., p. 782, 2009.

TENG, S.S.; CHENG, B.H. Formation of pyrochlorophylls and their derivatives in spinach leaves during heating. **Food Chemistry**,

v. 65, p. 367-373, 1999.

TOIVONEN, P.M.A.; SWEENEY, M. Differences in chlorophyll loss at 130C for two broccoli (Brassica oleracea L.) cultivars associated with antioxidant enzyme activities. **Journal Agricultural Food Chemistry**, v. 46, p. 20-24, 1998.

YUAN, G.; SUN, B.; YUAN, Y.; WANG, Q. Effect of 1-methylcyclopropene on shelf life, visual quality, antioxidant enzymes and health - promoting compounds in broccoli florets. **Food Chemistry**, v. 118, p. 774-781, 2009.

CHAPTER 4

CHLOROPHYLL AND CAROTENOID CONTENT IN LETTUCE PLANTS FROM SEEDS TREATED WITH DIFFERENT COLOURS

Letícia Winke Dias

André Pich Brunes

Andrea Bicca Noguez Martins

Olericulture plays an important role in agriculture and the economy in Brazil (GOMES et al., 2016), and lettuce is one of the most important vegetables. It is a leafy vegetable belonging to the Asteraceae family. It originated in Asia and was brought to Brazil by the Portuguese in the 16th century. It is one of the most consumed vegetables in Brazil due to its low cost, pleasant flavour and easy cultivation, making it easy to grow all year round with varieties adapted to the climate of each region (SILVA et al., 2015). Amongst the different varieties, we highlight the smooth, American, romaine, curly and purple lettuce (PHILIPPI et al., 2006).

Physiologically, it is a species that produces positive photoblastic seeds, which means that its seeds only germinate, or germinate in a higher percentage, in the presence of light. Therefore, due to the domestication process and large-scale production, many cultivars have already lost this photoblastism and have become neutral photoblasts (CASTRO et al., 2005). Plants perceive the quality of light through phytochrome, which is a class of photoreceptor pigments.

Priming is characterised by a pre-treatment carried out on the seeds, using different methods that can favour uniform germination and seedling emergence, due to the control of the amount of water available to the seed in the early stages of germination, so it will not allow root emission (USDA, 2014). It can be considered a common practice in the seed market when it is associated with the vegetable seed pelleting process.

Since the 1990s, the seed coating process has evolved so that today it is at the top of the seed industry, due to concerns about environmental protection and occupational safety, as well as precision sowing (SANTOS, 2016).

One of the most important plant components are pigments, which are organic compounds capable of absorbing electromagnetic radiation in the 400 to 700 nm range (KLUGE et al, 2015). They are located inside chloroplasts and the main groups of pigments found in plants are chlorophylls (such as chlorophylls a and b), carotenoids (such as p-carotene, lutein, violaxanthin and neoxanthin) and flavonoids (such as flavones, flavonols, isoflavones and anthocyanins).

Leafy vegetables contain various types of photosynthetic

pigments, which are chlorophylls and carotenoids (KIMURA & RODRIGUEZ-AMAYA, 2002). The levels of these pigments are used to estimate the photosynthetic potential of plants, due to their direct link to the absorption and transfer of light energy and to growth and adaptation to different environments. A plant with a high chlorophyll content is capable of achieving higher photosynthetic rates due to its potential value for capturing "quanta" in a unit of time (CHAPPELLE & KIM, 1992).

In this context, the aim of this study was to assess the levels of total chlorophylls and carotenoids in lettuce seeds pelleted with and without priming with different colours.

The work was carried out using lettuce seeds of the Regina cultivar from the same batch. The experiment was conducted from January to April 2017.

The experimental design was entirely randomised, in an AxB factorial scheme (Factor A: with priming and without priming, and Factor B: colourings (yellow, blue, orange, green and red), with four replicates.

Germination was carried out with 200 seeds, with four repetitions of 50 seeds for each treatment, distributed in plastic gerboxes (11.0 x 11.0 x 3.5cm) on two sheets of blotting paper moistened with a quantity of water equivalent to 2.5 times the weight of the dry paper, at 20°C, according to the criteria established by the Rules for Seed Analysis (Brazil, 2009). Samples were taken at the time of the lettuce seed germination test.

The variables assessed were:

Total chlorophyll, chlorophyll a, chlorophyll b and total carotenoid content - at seven days after sowing, for the germination test (BRASIL, 2009), 1g samples of lettuce seedlings were macerated in a mortar in the presence of 5 mL of 80% (v/v) acetone. The material was centrifuged at 4000±5000g for 10 minutes and the supernatant was transferred to a 25 mL volumetric flask and topped up with 80% (v/v) acetone. The absorbance of the solution was obtained by spectrophotometry at 647, 663 and 470nm. The total chlorophyll, 'a', 'b' and total carotenoid contents were calculated using equations (eq. 1, 2, 3 and 4) established by Lichtenthaler (1987):

Total Chl = 7.15(A663) + 18.71 (A647) (eq. 1)
Chi 'a - 12.25 (A663) - 2.79 (A647) (eq. 2)
Chl **'b - 21.50 (A647) -** 5.10 (A663) (eq. 3)
Total carotenoids = [1000(A470) -1**.82Chl'a'- 85.02 Chl'b'**] /198 (eq. 4)

The results were expressed in pg μg-1 of fresh matter (MF).

The data obtained was analysed for normality and then submitted to ANOVA analysis of variance. The qualitative factor was analysed by mean comparisons using the Duncan test at 5% probability. The statistical procedure used the R programme, version 3.1.1, and the "agricolae" data package (MENDIBURU, 2014; R CORE TEAM, 2014).

Table 1 shows that there was a significant interaction between the treatments colour, priming, colours and priming for the variables total chlorophyll, chlorophyll 'a', chlorophyll 'b' *and* total carotenoids.

Table 1. Analysis of variance (ANOVA) of lettuce seed treatments with and without priming and pelleted with different colours for the variables total chlorophyll, chlorophyll 'a', chlorophyll 'b' and total carotenoids.

	Total chlorophyll	Chlorophyll 'a'	Chloroph yll 'b'	Total Carotenoids
Colour	-0	-0	-0	-0
priming	$8{,}36\times10^{-7}$	$1{,}40\times10^{-7}$	$1{,}53\times10^{-5}$	$5{,}83\times10^{-8}$
colour. *priming*	-0	-0	-0	-0

The priming treatment resulted in a higher total chlorophyll content in the lettuce seedlings when it came from seeds pelleted with yellow, orange, green and red colours, while the treatments without priming were only superior in the blue colour (Table 2). The yellow and red colours showed the highest total chlorophyll content in the priming treatments, followed by the orange and green colours, which did not differ from each other, and blue, which showed the worst result. For the treatments without priming, the blue colour had the highest total chlorophyll content and the yellow, orange and red colours had the lowest results.

Similarly for chlorophyll 'a' and 'b', the priming treatments showed higher levels of these compounds in the yellow, orange, green and red colours, with only the blue colour being higher in the absence of priming (Table 2). In both compounds, chlorophyll 'a' and 'b', among the *priming* treatments the yellow colour was higher than the others and the lowest levels were found in the seedlings from pelleted seeds with the blue colour. In the treatments without *priming*, the blue colour was higher than the other colours.

According to Taiz and Zeiger (2013), carotenoids are pigments that act as photoprotectants by rapidly quenching the excited states of chlorophyll a and are generally closely associated with the protein pigments of reaction centres and antennae.

In this study, total carotenoid content in the absence of *priming*, seedlings from pelleted seeds coloured blue were the ones with the highest content of these compounds. In the case of the other coloured pelleted seeds, the highest accumulation of these compounds occurred in the presence of *priming* (Table 2).

Regarding the results found for 'a', 'b' and total chlorophylls, among the treatments with *priming, the* yellow colour was the one that resulted in the highest total carotenoid content, with the blue colour being inferior to the others. Among the treatments without *priming*, blue colouring was superior to the others, with yellow colouring having the worst result.

Table 2. Total chlorophyll, chlorophyll 'a', chlorophyll 'b' and total carotenoid contents from seeds treated with and without priming and pelleted with different colours.

colour / priming	total chlorophyll		chlorophyll 'a		chlorophyll 'b'		total carotenoids	
	mg g^{-1} of dry matter							
	with	without	with	without	with	without	with	without
Yellow	*a 6.21 A ᵅ	d 2.57 B	to 3.46 A	d 1.64 B	to 2.74 A	b 0.93 D	to 395.2 A	b 152.10 D
Blue	c 2.81 B	to 6.88 A	c 1.63 c B	to 3.83 A	b 1.18 D	to 3.05 A	b 157.2 D	to 428.4 A
Orange	b 4.08 A	c 3.10 B	b 2.30 A	bc 1.93 B	to 1.77 B	b 1.17 C	a 265.4 B	b 186.1 C
Green	b 3.82 A	b 3.44 B	b 2.34 A	b 2.06 B	at 1.48 C	to 1.38 B	to 236.9 C	b 212.4 B
Red	b 3.97 A	bc 3.36B	b 2.46 A	c 1.83 B	at 1.50 C	to 1.53 B	to 241.9 C	b 222.5 B
Average	4,18	3,87	2,44	2,26	1,74	1,61	262,1	240,3
C.V. (%)	3,90		3,57		4,62		4,01	

Averages followed by the same letter in the line do not differ according to the Tukey test ($p < 0.05$).
"Averages followed by the same letter in the column do not differ according to the Tukey test (P 0.05).

Studies show that chlorophylls are pigments that reflect the colour green and are entirely associated with the potential for photosynthetic activity. High photosynthetic efficiency can lead to an increase in agricultural productivity, and this relationship is directly related to the utilisation of the radiation available from these pigments (SILVA et al, 2014). Since chlorophyll a (Chl a) is present in all organisms that carry out photosynthesis, it is the pigment used to carry out the photochemical phase, while the other pigments help to absorb light and transfer energy to the reaction centres and are called accessory pigments (TAIZ & ZEIGER, 2013). It is possible that carotenoids act as antioxidants against the acceleration of ethylene production and respiration (VALKO et al.,

2004), protecting, along with the increase in the content of carotenoids in the plant.
chlorophylls and, consequently, favouring the maintenance of the greenish colour (ZAICOVSKI et al., 2008).

We can conclude that, in terms of total chlorophyll, chlorophyll 'a', chlorophyll 'b' and total carotenoid content, the colourings from the yellow and blue pellets generally had a higher amount of these compounds than the other colourings.

References

BRAZIL. Ministry of Agriculture, Livestock and Supply. **Rules for analysing seeds.** Ministry of Agriculture, Livestock and Supply. Secretariat for Agricultural Defence. Brasília, DF: Mapa/ACS, 2009. 395p.

CASTRO, P.R.C.; KLUGE, R.A.; PERES, L.E.P. **Manual de Plant Physiology: Theory and Practice.** Piracicaba: Editora Agronômica Ceres. 2005. 650 p.

CHAPPELLE, E.W.; KIM, M.S. Ratio analysis of reflectance spectra (RARS): an algorithm for a remote estimation of the concentrations of chlorophyll A, chlorophyll B, and carotenoids in soybean leaves. **Remote Sensing of Environment,** v. 39, p. 239-247, 1992.

GOMES, I.S.; OLIVEIRA, C.S.; PELOSI, A.P.; PACHECO, L.C.P. S.; BENETT, C.G.S.; BENETT, K.S.S. Evaluation of physiological quality of lettuce and rocket salad seeds in the presence of purple nuts edget extract. **African Journal of Agricultural Research**, v. 11, n.21, p. 1887-1893, 2016.

KIMURA, M.; RODRIGUEZ-AMAYA, D.B.A. scheme for obtaining standards and HPLC quantification of leafy vegetable carotenoid. **Food Chemistry**, v. 78, p. 389-398, 2002.

KLUGE, R.A.; TEZOTTO-ULIANA, J.V.; DA SILVA, P.P.M. Physiological and Environmental Aspects of Photosynthesis. **Rev. Virtual de Química**, v. 7, n. 1, p. 56-73, 2015.

LICHTENTHALER, H.K. **Chlorophylls and carotenoids: pigments of photosynthetic biomembranes. In: Packer, L., Douce, R. (Eds.).** Methods inenzimology, Academic Press, v.148, p. 350-381, 1987.

MENDIBURU, F. **Agricolae: Statistical Procedures for Agricultural Research. R package version 1.2-0**. 2014. http://CRAN.R-project.org/package=agricolae

PHILIPPI, S.T. **Nutrição e Técnica Dietética**. 2ª revised and updated edition. Barueri, SP: Editora Manole, 2006.

R CORE TEAM. **R: A language and environment for statistical computing.** R Foundation for Statistical Computing, Vienna, Austria. 2014. Available at: http://www.R-project.org/.

SANTOS, S.R.G. Forest Seed Pelletisation in Brazil: An Update. **Revista Floresta e Ambiente**, v. 23, n. 2, p. 286294, 2016.

SILVA, M.R.P.; PINHEIRO, F.C.; PAULA, M.T.; PRIGOL, M. Parasitological evaluation of lettuce (Lactuca sativa) sold in a municipality on the western border, Rio Grande do Sul, Brazil. **Revista de Patologia Tropical**, v. 44, n. 2, p. 163-169, 2015.

SILVA, M.A; SANTOS, C.M.; VITORINO, H.S.; RHEIN, A.L.F. Photosynthetic pigments and spad index as descriptors for

intensity of water deficiency stress in sugarcane. **Bioscience Journal**, v. 30, n. 1, p. 173-181, 2014.

TAIZ, L.; ZEIGER, E. **Plant Physiology.** Trad. Armando Molina Divan Junior... [et al.] - 5. ed. Porto Alegre: Artmed. 2013. 918 p.

USDA **United States Department of Agriculture**. Available at: http://agclass.nal.usda.gov/mtwdk.exe?k=glossary&l=60&w=10684& n=1&s=5&t=2

VALKO, M., IZAKOVIC, M., MAZUR, M., RHODES, C.J., TELSER, J. Role of oxygen radicals in DNA damage and cancer incidence. **Molecular and Cellular Biochemistry**, v. 266, p. 37-56, 2004.

XUE, L., YANG, L. Deriving leaf chlorophyll content of green-leafy vegetables from hyperspectral reflectance. **ISPRS Journal of Photogrammetry and Remote Sensing**, v. 64, p. 97-106, 2009.

ZAICOVSKI, C.B. **Caracterização de frutas nativas da Região Sul da América do Sul quanto à presença de compostos bioativos, da atividade antioxidante e da atividade antiproliferativa frente à celular tumorais**, RS, Pelotas. 2008, 91f. Thesis (Doctorate in Agroindustrial Science and Technology) - Eliseu Maciel School of Agronomy, Federal University of Pelotas, Pelotas.

FSC
www.fsc.org
MIX
Papier | Fördert gute Waldnutzung
FSC® C083411

Kolibri 360 GmbH
Ferdinand-Jühlke-Straße 7
99095 Erfurt, Deutschland
produktsicherheit@kolibri360.de

Druck:
CPI Druckdienstleistungen GmbH
im Auftrag der
Kolibri 360 GmbH
Ferdinand-Jühlke-Str. 7
99095 Erfurt